ADDRESS

BEFORE

THE AMERICAN ASSOCIATION

FOR THE

ADVANCEMENT OF SCIENCE,

AUGUST, 1859.

BY

PROFESSOR ALEXIS CASWELL.

PUBLISHED FOR THE ASSOCIATION,

BY JOSEPH LOVERING.

PERMANENT SECRETARY.

1859.

ADDRESS

BEFORE

THE AMERICAN ASSOCIATION

FOR THE

ADVANCEMENT OF SCIENCE,

AUGUST, 1859.

BY

PROFESSOR ALEXIS CASWELL.

PUBLISHED FOR THE ASSOCIATION,

BY JOSEPH LOVERING,

PERMANENT SECRETARY.

1859.

CAMBRIDGE:
Allen and Farnham, Printers.

ADDRESS

OF

PROFESSOR ALEXIS CASWELL,

PRESIDENT OF THE AMERICAN ASSOCIATION FOR THE YEAR 1858,

ON RETIRING FROM THE DUTIES OF PRESIDENT.

MR. PRESIDENT, AND GENTLEMEN OF THE AMERICAN ASSOCIATION FOR THE ADVANCEMENT OF SCIENCE: —

IT sometimes happens that one finds himself in a position which requires an explanation, or even an apology. Such is my own case this evening. The duty which devolves upon me more properly belongs to another. Your Constitution provides for a Retiring Address from your President. He becomes, for the time being, your official adviser, — the exponent of your views. This is one of the objects contemplated in calling him to the presidency. Your Vice-President holds a somewhat different relation to the body, and may not, without seeming arrogance, take upon himself the special functions of his official superior.

My own predilections and sense of fitness would have precluded my appearance before you in this official capacity.

Hitherto, on these occasions, you have listened to the accredited representatives and leaders of American science, each of whom might well apply to himself the motto, "Quorum magna pars fui." I am but an humble laborer in the ranks. But I am here at your command, which, having for its object the advancement of science, is not lightly to be set aside. I should be unworthy of the confidence which you have reposed in me did I not endeavor to comply with your wishes in giving to our proceedings that completeness which the constitution contemplates. But in doing this, if I shall fail to offer any thing worthy of your attention, I beg you to bear in mind that it is not a failure of my own seeking. When I have done what I can, I shall take shelter in a divided responsibility.

Permitted to convene on this, our Thirteenth Annual Meeting, it is fit that we should, as a first duty, pay a passing tribute of respect to the memory of those who have left us. It is well that we should pause even in the grave and urgent pursuits of life to ponder on its brevity.

Since our last meeting, several who have been accustomed to participate in our proceedings, have ceased to be numbered with the living. Prof. Ira Young, of Dartmouth College; Prof. William M. Mather, of the University of Ohio, and at the time of his death the acting President; and Prof. Denison Olmsted, of Yale College, have been borne to their long homes. They were all widely known in Academic circles.

Prof. Young was born in the town of Lebanon, New Hampshire, May 23, 1801. His father was by trade a carpenter, which trade the son also followed till he had attained his majority. He early manifested mechanical ingenuity and an ardent love of knowledge. On arriving at the age of twenty-one, he commenced a course of study preparatory to entering college; which, with his diligence and ability, was soon completed. He entered Dartmouth College, and, having completed the academic course, graduated with distinction in 1828. Like many distinguished graduates of our colleges, he defrayed the expenses of his education mainly by the fruits of his own industry. He was soon called to the post of Tutor in the college, and after two years of service in that capacity, was elected to the Professorship of Natural Philosophy and Astronomy, which chair he filled with marked ability and success, till the time of his death, Sept. 13, 1858, — a period of twenty-five years. He was a man of exact knowledge, of great firmness, of inflexible integrity, and of exemplary Christian piety. He was well versed in all the subjects pertaining to his department, and was, in the opinion of those most familiar with him and most competent to judge, an admirable teacher. He died in the fifty-eighth year of his age.

Prof. Mather was a native of Middlesex county, in the State of Connecticut. The date of his birth has not fallen under my notice. His name appears in the list of graduates of the United States Military Academy, at West Point, for 1828. He entered the army, but in 1829 was appointed Assistant Professor of Chemistry, Mineralogy, and Geology in the Academy, and continued to discharge the duties of that post for a period of six years. He resigned his commission in the army in 1836, and was elected to the professorship of Chemistry in the University of Louisiana, which place he held but for a short period. He was connected in succession with the geological surveys of the States of New York, Ohio, and Kentucky. His final report on the geology of the first district in New York was published in a large quarto volume in 1823. This is regarded as his most important original work, and in the opinion of competent judges, bears "honorable testimony to his ability and accuracy as an observer in this department of nature." * He also published valuable reports of his labors in connection with the surveys of Ohio and Kentucky. From 1847 to the time of his death, Feb. 29, 1859, he was Professor of Natural Sciences in the University of Ohio, and for some time previous to his death acting President of that institution. He was a valuable contributor for many years to the American Journal

* American Journal of Science, Vol. XXVII. p. 452.

of Science, in the departments of Chemistry, Mineralogy, and Geology.

For the facts connected with the life of Prof. Olmsted I am indebted mainly to a discourse commemorative of his life and services, delivered in the college chapel at New Haven, shortly after his death, by President Woolsey.

From this it appears that he was born in East Hartford on the 18th of June, 1791. He was a descendant of James Olmsted, one of the first settlers of the colony of Connecticut. Bereaved of his father while but an infant, his youthful training was under the direction of his pious and most excellent mother. He gave early indications of a love of books, and diligently availed himself of such slender means of instruction as the common schools of the place and the moderate resources of his family could afford. With the aid and encouragement of a few friends who became interested in his youthful promise, he prepared himself for admission to Yale College in 1809, where he was graduated with distinction in 1813. During his collegiate course he indicated no marked preference for any particular branch of study. If there was any department in which he might be said to be distinguished, it appears to have been that of literature. "He excelled in writing," says President Woolsey, "and cultivated a taste for belles-lettres and poetry." On leaving college, he taught for a while a select school for boys, and was then appointed a tutor in the institution; where, in addition to the duties of his tutorship, he commenced the study of theology, under the supervision of that distinguished teacher and theologian, Dr. Timothy Dwight. During this same year, Mr. Olmsted was appointed Professor of Chemistry in the University of North Carolina, the sciences of geology and mineralogy being included in the professorship. While connected with that university he was invited by the State to undertake a survey of its geology and mineral treasures. A report of these labors was published in 1824–5, under the direction of the State Board of Agriculture. After a period of seven years' service at Chapel Hill, he accepted, in

1825, the appointment of Professor of Mathematics, Natural Philosophy, and Astronomy in his Alma Mater. For thirty-four years, with the utmost diligence and fidelity, he fulfilled the responsible duties of this station, with the exception that the department of Mathematics was after a few years constituted a separate professorship, and placed in other hands. It may be remarked in this connection, that his tastes and his early professional studies, did not lead him to the cultivation of the higher branches of the Mathematics. Throughout the long period of his professional life he sustained the reputation of an excellent teacher, specially apt in the clear and skilful presentation of whatever subject he had occasion to treat before his classes. He was the author of several text-books for colleges, which have passed through many editions. He also gave to the public several popular compends of science, designed for more general use. His text-books for colleges were his Natural Philosophy and his Astronomy. In both of these works he has restricted himself to the more elementary portions of the sciences, and has employed exclusively the geometrical methods in distinction from the analytical. It seems to have been with him through life a leading object to make science popular, and this perhaps determined him in the particular mode of presenting it, deeming the geometrical method, of the two, the more readily understood. However true this may be, — and for beginners it undoubtedly is true, — the calculus, since the time of Newton, has been the great instrument for developing, in all their extent, the physico-mathematical sciences, and cannot be too strongly commended to those who would prepare themselves for independent and aggressive research.

In addition to his scientific and more popular books, Prof. Olmsted was a very frequent contributor to the pages of the American Journal of Science and Art. He was also an early and efficient friend of popular education, and was among the first to urge the importance of normal schools for the education of teachers.

The extraordinary shower of meteors during the night of

Nov. 12, 1833, drew the attention of Prof. Olmsted to the investigation of that subject. After a very laborious inquiry, both historical and scientific, he came to the conclusion that these meteors were portions of the extreme parts of a nebulous body which revolves around the sun interior to that of the earth, but little inclined to the plane of the ecliptic; having its aphelion near the earth's path, and having a periodic time of 182 days nearly.* These researches attracted the attention of the public at home and abroad, and received the commendation of several of the most eminent astronomers of Europe. The explanation here given unites, perhaps, a larger number of suffrages than any other, and yet is hardly to be considered as one of the established facts of science.

The elaborate memoir of Prof. Olmsted, entitled, "On the Recent Secular Period of the Aurora Borealis," published by the Smithsonian Institution, may justly be regarded as one of the most valuable papers which have been given to the public on the perplexed subject of the Northern Lights.

Few men have been called to drink more deeply of the bitter cup of affliction than our deceased friend and associate. Between the years of 1844 and 1852, he buried four dutiful and promising sons,—all graduates of Yale College. Their mother had died before them. These heavy bereavements he bore with singular Christian fortitude and resignation. As a citizen, teacher, and friend, he was universally beloved. I should fail in doing justice to his memory did I not quote a single sentence, illustrative of his character, written a few days before his discease. "In view of the uncertain issue of this sickness," he writes, "I desire humbly to cast myself upon God, humbly to implore his forgiveness of my sins through Jesus Christ, and to express a cheerful hope that, should I be called away, it will be to my heavenly Father's house."

We now pass to other considerations appropriate to this occasion.

* Woolsey's Discourse, p. 17.

It is not expected that this Address will take the form of a scientific paper. It embraces a wider scope and more varied topics. And yet it is hoped that it will not, on that account, be the less pertinent to the general purposes of our meeting.

The object of this Association is the advancement of science. Here two things are implied: first, that science is an object worthy of our pursuit; and, secondly, that combined efforts are necessary to its progress. A few moments' consideration of these points may not be without interest.

1. To say that science is worthy of our pursuit, may at first seem to be a mere waste of words, — the statement of a proposition to which every one assents. But what do we mean by science? Is it the method of converting the products of the fields and forests into articles of commercial value? Does it specially embrace those labor-saving machines by which one man does the work of twenty or an hundred? Is it chiefly found in the immense accumulation of mechanical force by which ponderous masses of iron and stone and timber are handled and tossed about as if they were mere playthings in the hands of giants? Certain it is that science has largely contributed to these results. But results like these by no means limit the value and range of science. These are examples of its immediate ~~ability.~~ But it has claims to our consideration beyond its immediate utility. The high and very just appreciation of what is termed practical and useful has, no doubt, in many instances, operated to the injury of what is termed speculative and theoretical. Nor is this singular. The random speculations and the fallacious theories with which the world abounds, under the pretence of improvements and progress, have contributed in the general mind to depreciate all scientific results, I will not say, which are not susceptible of practicable applications (for of this we cannot judge), but which have not yet been applied to practice. There is everywhere a pressing demand for what is practical, and often a narrow utilitarianism, which looks with suspicion, and sometimes with contempt, upon the profoundest speculations of

science, because they do not pay back to the laborer the wages of his hire in marketable value. The scientist, no less than the severest utilitarian, rejoices in every contribution of science to the arts, but he does not admit that the whole value of science is to be measured by any present applications. He puts in a demurrer to the conclusion that those portions of it are useless of which we do not now see the utility. The use may be beyond our present sphere of vision, or, if coming within our cognizance, it may not admit of comparison with any standard of measure. Unlike the precious gem, which has an exchangeable but no intrinsic value, science bears no price in the market. It transcends all ideas of comparison and exchange. Its high utility lies in the breadth and dignity and sublime grandeur which it gives to the human mind.

In the succession of geological strata, buried deep in the earth, with here and there an upheaval which brings them into view, as in the computed periods of astronomical perturbations, what vast cycles of time are brought within the grasp of the understanding? It is not merely the naked, isolated conception of an immense duration, but a distinct and orderly succession of events, which makes up the physical history of that duration.

In the narrow sense of utility, it may be of little use to know what relation or fellowship exists between the two companions of a minute double star, — minute to us, but of vast intrinsic splendor; — but what educated mind can be indifferent to the fact that the identical law of gravity which determines the fall of an apple, binds those stars to each other, and compels them, through vast cycles of time and almost boundless space, to trace their elliptic orbits? That, to our apprehension, is the grandest law of nature. It controls the material universe, and heralds from sphere to sphere, in endless succession, the order and harmony which emanate from the Creator's will. It is thus, under the guidance of science, that our conceptions of law and order, of space and time, lead us to the Infinite. We look at the starry hosts, — myriads though they be, — and ask, are there not more? Is it in the lowest degree probable that the material world is bounded by the utmost limits of our telescopic vision? The universal dominion of the same law, the final stability which prevails over inevitable and endless perturbations, mark the universe as a system of exquisite and most perfect mechanism. It is the work of one Intelligence, and that one Intelligence is infinite. I cannot but think that the profoundest knowledge of nature leads, logically, to this conclusion. And this is one of the grandest utilities of science. It raises the human mind to the intelligent contemplation of the Deity.

But this is not all. The ministry of science stops not here. I go further and maintain that a profound knowledge of nature, in her laws and adaptations and vastness, puts us in possession of the truest and best means of weighing the evidence of a divine revelation, and suggests many analogies in its defence. I shall not hesitate, in this presence, to give utterance to my own profound conviction that all true science is in harmony with the Bible, rightly interpreted. Any seeming discrepancy which baffles the resources of ingenuity to reconcile is but the varying ripple on the mighty swell of the ocean, whose exact form no power of analysis can express, and no skill of the pencil sketch. It is but the fleecy cloud which dims a little patch of the deep blue sky, or the smallest visible spot which can be discerned on the broad, blazing disc of the sun, which would not be seen at all but for the splendor of its surroundings.

Science is not unfriendly to religion. True it is that some men, distinguished for science, have appeared to disregard the obligations of religion, and others have equally disregarded the obligations of morality. But science cannot be held responsible, nor ought it to suffer reproach for delinquencies like these. Copernicus was a grave and devoted priest of religion, giving his energies, first to the cure of souls, then to the ministrations of science. Kepler, it is said, seldom entered upon an important investigation without first seeking, by prayer, the wisdom which is from above. Newton was a devout Christian, scarcely less distinguished for his piety than for his unrivalled power of mathematical analysis and his rare skill in experimental research. Examples like these, of which the annals of philosophy are richly stored, while they prove nothing in respect to religion, may well serve to exonerate science from the charge of infidel tendencies.

I have adverted to this topic for the purpose of glancing at the higher ends of science, and specially for the purpose of relieving the fears of any who may look upon the bold and fearless advance of physical investigations as ominous of evil to the interests of religion. We participate in no such fear. We wish explicitly to exonerate this Association from all suspicion of undermining, or in any manner weakening, the foundations of that faith which an apostle says was once delivered to the saints. We cannot admit the opinion, that any progress in science will ever operate to the disparagement of that devout homage which we all owe to Him in whose hand our breath is and whose are all our ways. Science, on the contrary, lends its sanction and adds the weight of its authority to the sublime teachings of revelation. Such are its ends, over and beyond its immediate practical utilities, and such are its claims upon our service.

2. We may now say, that combined efforts are necessary to its further progress.

The mathematical sciences grow by a development from within; the physical mainly by accretions from without. To the progress of pure analysis it might seem that there was little opportunity for the coöperation of different minds, — that one master-spirit must lead the way, and all others must follow. The coöperation here is not by simultaneous, so much as by successive, labors. The ancient geometers laid the foundation; the modern have built the superstructure. Each successive laborer has resumed the work where his predecessor left it. Each has added his link to the golden chain. And yet even here so varied and distinct are the fields of analytical research, that the demand for labor is always fully equal to the supply.

The physical sciences advance by observation, by the accumulation of fact, by experimental research and induction. Here the simultaneous coöperation of many laborers is often indispensable. In the science of Meteorology, for example (if in its present state it can properly be called a science), what an accumulation of observations must be made, simultaneous, widely extended, and long continued, before the work of discussion and critical analysis can hope to develop minutely the laws of atmospheric change. These changes, however seemingly capricious, are doubtless as much in accordance with fixed laws as the planetary orbits. But whether the complicated elements which enter into them ever can be grasped and reduced to form, so as to give to the science that completeness which belongs to Astronomy, may well be doubted. That results most important to agriculture, to commerce, and to the sanitary condition of mankind may be reached, is no longer problematical. It may be regarded as one of the certainties of future progress. But these results must be based upon the comparison and discussion of a wide range of observation. Much is to be hoped in this respect from the system of observations organized by the Smithsonian Institution in this country, in connection with similar organizations in England, France, and Germany. The same extended coöperation is required for the advancement of Geology, Mineralogy, Terrestrial Magnetism, Botany, Zoölogy, and indeed of all the departments of Natural History.

What we want, then, and what we aim at, in this Association, is to render the scientific resources of the country available. The coöperative agencies are numberless. Every eye that can see, every ear that can hear, every hand that can record, and every intellect that can analyze and combine, may contribute to the grand result. We invite the widest coöperation. The wisdom of this body will point out the most productive fields of labor, and, as far as possible, suggest the best methods of research.

But in order to gather around us, and bring into active co-operation the scientific talent of the country, we must work with a true devotion to science. By our practical wisdom, our friendly criticism, our impartial justice, our magnanimity and self-sacrifice, we must show that we deserve the confidence of an enlightened public.

In glancing at the field before us, I must limit my remarks to a single department, — that of Astronomy, — to which my own attention has been more particularly given.

It has long been the boast of Astronomy, that it is the oldest and most perfect of the sciences. But how can we predicate perfection of a science that is in a state of rapid advancement? Astronomy at the present time must be regarded as one of the most progressive of the sciences. One or two illustrations of its progress will not be out of place. Tycho Brahè was a great astronomer. His magnificent observatory has a world wide celebrity. And yet it was an observatory without a telescope. Compare the pinnules and coarse graduations of his instruments with the refinements of modern telescopes, circles, and micrometers, and you have an impressive idea of progress. By a laborious comparison of Tycho's observations, Kepler arrived at the conclusion that the planets revolve in elliptical orbits. By the law of gravitation it is now shown that those elliptical orbits are subject to perpetual perturbations, and the minute changes which they must severally undergo for ages to come are pointed out with almost unerring precision. These perturbations themselves become, in the hands of the physical astronomer, unequivocal exponents of the condition of the system. They open up new vistas into space, through which his far-reaching analysis catches the gleams of coming light. The discovery of the planet Neptune will ever constitute a brilliant epoch in the annals of Astronomy, and place the names of Leverrier and Adams in the foremost rank of geometers. The planet Uranus was known to be subject to slight irregularities of motion. Conjecture had for some time assigned the cause to be the attraction of an unknown planet. Analysis in the hands of these geometers confirmed the existence of such planet, and pointed to its place in the far off depths of ether. And on turning the telescope to the spot, behold there it was. Such an achievement belongs to the ninteenth century, and was scarcely possible at an earlier period. Without accurate measurements the irregularities of Uranus could not have been detected; without that analysis, itself one of the noblest monuments of the human intellect, the place of the unknown planet could not have been conjectured.

As a further illustration, let me refer to the recent labors of Hansen in perfecting the Lunar Tables. The irregularities of the moon's motions, owing to the perturbation of the sun and planets, have been the subject of perplexity and anxious inquiry with astronomers from the time of Hipparchus. No tables could be constructed which would assign the moon's place with accuracy for any long period. So great was the difficulty of the problem that Euler despaired of ever being able to reduce the error in the computed place of that body, to any very narrow limits. Hansen undertook the reëxamination of the Lunar Theory, as based upon the best modern observations, and particularly upon those of the Greenwich Observatory. His profound knowledge of the theory of perturbations enabled him to detect two inequalities of considerable magnitude, due to the action of Venus; which being incorporated in the tables relieves the subject almost entirely of embarrassment, and assigns the moon's place for any epoch with all the precision required by the present state of the science.

With these improved tables, Professor Airy has been enabled to settle the long contested epochs of several memorable ancient eclipses; as, for instance, those of Thales, and Agathocles, and that at Larissa; and has thereby established several important points in ancient chronology beyond all reasonable doubt. A curious fact connected with one of these eclipses shows in what manner the light of science reflected backward sometimes illustrates the dark points of history in the remote past. I am tempted to refer to it. It is stated by Diodorus

Siculus that the fleet of Agathocles, the tyrant of Sicily, was blockaded in the harbor of Syracuse, by the Carthaginians, with whom he was then at war. Shut up within narrow limits, and pressed on every side, he resolved on the bold and memorable project, then for the first time contemplated in the history of Carthaginian warfare, of "carrying the war into Africa." Under cover of the night, aided by many ingenious stratagems, he escaped from the harbor, and sailed for the African coast, pursued by the blockading fleet. Six days after his escape, having eluded the pursuit of the enemy, he landed at the "Quarries," burnt his ships, sacrificed to the gods, and commenced a campaign of conquest and slaughter which has few parallels in the history of warfare. An interesting account of it is given by Grote. On the day after the escape of Agathocles from Syracuse, "there was such an eclipse of the sun," says Diodorus, "that the day wholly put on the appearance of night, stars being seen everywhere." * History has left it uncertain when this memorable event occurred; and it has been made a critical question whether Agathocles passed round on the north or on the south side of the island. Singular as it may seem, the astronomical researches of Hansen and Airy have shown with more than historic certainty, that on the 15th of August, 309 before Christ, at six o'clock and forty minutes in the morning, Greenwich mean time, Agathocles witnessed a total eclipse of the sun, and must have been on the *south side of Sicily.*

But in the present advanced state of astronomical science, what, it may be asked, remains for astronomers to do? What new and noble achievements invite their toil? It would not be easy, perhaps not possible, to enumerate all the unsolved problems of astronomy. Several subjects of great interest may, however, be mentioned, which await further investigation.

(1) The true dimensions of the solar system, however much has been done in that direction, are by no means settled

* As quoted by Prof. Airy, Phil. Trans. Vol. LXXI. p. 187.

with that exactness which precludes the desire of closer approximation. The great primary problem, next to the size of the earth, is the distance of the sun. This has been pronounced by Airy to be the grandest problem in astronomy. I will not here detail the methods by which any error in the present determination may be eliminated or reduced. I shall refer to it again. With any correction in the sun's distance, all the planetary distances will take new value; and with these corrected distances, it may be necessary to reconstruct the tables of all the older planets. Then there are the asteroids, now numbering, I think, fifty-five (55). A vast amount of labor is necessary to determine their exact orbits. Their mutual perturbations will demand and receive attention. Nor will the possibility of their ever having ~~found~~ a single body, broken in pieces by some violent shock of nature, be permitted to pass without severe scrutiny; nor the other possibility of two or more of them being, by their mutual attractions, drawn together and forming a double planet. The physical constitution of the planetary bodies, and especially that of the sun, presents questions of great interest, which it would be unphilosophical to despair of being able to solve. The solar spots, of which the number has been very large during the present season, deserve, it would seem, to be scrutinized with the highest powers of the telescope. The present theories respecting them are, by no means, so satisfactory as to preclude further and more thorough investigation. Such an inquiry would probably determine more exactly than is now known, the time of the sun's rotation on his own axis. Whether it would throw any light upon the cause of their greatest and least abundance in periods of about eleven years, as discovered by Schwabè, time must determine. Do these spots occasion any sensible change in the solar radiation, and consequent change in the temperature and productions of the earth? Do they exist equally in all parts of the equatorial region? Or are there special localities for the development of those immense tornadoes whose ravages we see in the blackened face of the sun?

(2) Leaving our own planetary system, when, let us ask, will the department of cometary astronomy be exhausted? At any moment, these strange visitors may burst upon us in mild or terrific splendor, as the case may chance to be. What variety of form do they present, what tenuity and what rapid and prodigious diffusion of luminous matter, driven off as if by the impulse of solar light? How soon the earth will come into collision with some one or more of the millions which are supposed to exist, and they swallow us, or we swallow them, none can tell. It requires evidently a special corps of observers, a scientific vanguard, to watch their movements and ward off. But, gravely, what are the internal forces which produce such immense diffusion of cometary matter in so short a time? Do the portions of the tail driven off many millions of miles from the nucleus, ever return to it? Or do they remain suspended in mid ether, in a condition to be gathered up by the next one which sweeps along the sky?

(3) Sidereal astronomy is itself a subject of boundless extent. With the aid of instruments of great refinement, many thousands of stars have been observed and their places determined with great precision. But this is not, like the computation of a table of logarithms, a work for all time. At no very distant epochs the entire sidereal heavens must be reëxamined. The stars are said to be *fixed*, but reëxamination shows that very many of them are *not* fixed. They have motions of their own — some greater, some less. What may be the result of the comparison of places, accurately determined for very distant epochs, we cannot predict. If the catalogue of Hipparchus, transmitted to us by Ptolemy, containing upwards of a thousand stars, could have been based upon the precise determination of modern instruments, we might now, looking upon the exact changes of two thousand years, be able to predict the further changes for many thousands of years to come. We might have under our immediate inspection, as in the case of the planet Neptune, a small portion of the orbit, which would enable us to describe the whole. In that case,

to the contemplation of the astronomer, how grand and solemn would now be the order and movement of the starry hosts! To temporary inspection they are all fixed; — to the eye of a thousand years *all are in motion*, — all pursuing their appointed orbits to be circled only in millions of years.

(4) How much is yet to be known of the three thousand and upward of double stars already catalogued! The periodic times, the positions and forms of the orbits of several, have already been proximately determined. In one instance, at least, that of Gamma Virginis, we have seen one star occulted by the other. It is said that in 1836 no telescope in Europe could separate the two stars. They are now quite wide apart. Continued attention to this class of objects will undoubtedly reward the inquiry with most interesting results.

(5) But what shall we say of variable and periodic, of new and lost stars? What a boundless field of observation, inquiry, and conjecture, does this subject open before us! — a field scarcely yet entered upon, but one which is, no doubt, destined to yield the richest harvest to the labors of future astronomers. What are the physical conditions which determine Algol, in the short space of three and a half hours, to change from a star of the second, to one of the fourth, magnitude, remain at its minimum brightness fifteen minutes, and then in three and a half hours more return to its former magnitude, and go through the same changes continuously in a period of little more than two days? What shall we think of Mira Ceti, which remains completely invisible to the naked eye for five months, then increases for three months, when it attains to the brightness of a star of the second magnitude; retaining this brightness for about two weeks, it again decreases for three months and disappears as before? It occupies about eleven months in passing through these changes. A still greater wonder is Eta Argus, which changes at irregular intervals from a star of the fourth, to one of the first magnitude, second only to Sirius in brightness. The whole number of variable stars which have

been subjected to a more or less careful examination, is from thirty to forty. Hind has remarked that the prevailing color among them is of a *reddish hue;* and in several instances, at least, during the period of least brightness they are surrounded by a hazy or nebulous light. These facts seem to point to some common characteristic. They indicate like physical conditions. "The whole subject of variable stars," says Herschel, "is a branch of practical astronomy which has been too little followed up,—it holds out a sure promise of rich discovery." It is also a branch in which amateur astronomers, with moderate instrumental means, or even with good eyes alone, can labor to advantage.

(6.) But we have not yet reached the widest field of inquiry. With telescopes of good definition and great power of illumination, the nebulæ present objects of interest which are hardly second to any others. The results furnished by the great reflector of Lord Rosse encourage the hope of most decided progress in this branch of astronomy. Whether the nebulæ are all groups of stars capable of being resolved, or in some cases immense diffusion of luminous matter in the slow process of condensation,—worlds undergoing a formative process,—they are alike wonderful to contemplate.

Such is a mere glance at the unexplored regions of Astronomy.

It is now time to ask, What may American astronomers be justly expected to do in advancing this, the noblest of sciences? We may best answer this question by looking at what they have done. In adverting to this topic it will be my endeavor not to transcend the limits of modesty and truth. I think it will appear from the educational and scientific history of the country, that scientific investigations, and especially astronomical pursuits, have always been held in high estimation among us. In proof of this I may allude to observations made at the last two transits of Venus. It is well known that Prof. Winthrop of Harvard College, visited Newfoundland in 1761, to observe the transit which occurred on the 5th of June of that year. It is not, perhaps, so well known that the "Province Sloop" was fitted out at the public expense to convey himself and his party to the place of observation. Elaborate preparations were made in different parts of the country to observe the transit of June 3d, 1769. Private munificence in the purchase of suitable instruments, and the coöperation of public bodies, bear honorable testimony to the enlightened zeal which animated the friends of science at that time. The observations of Rittenhouse and his associates, in Philadelphia, of Winthrop in Cambridge, and West in Providence, rendered important aid in determining the problem of the sun's parallax.

But while an honorable appreciation of astronomical science has been perpetuated among us, strange as it may seem, until within a very recent period, we have wanted the instrumental means and appliances necessary to prosecute it on independent grounds. So recently as 1840 we had done almost nothing in the way of astronomical ~~experiments.~~ There were at that time in possession of the coast survey, several fine instruments for the determination of latitude, for transits and for moon-culminations; and the reports of that department show how well they were used. Yale College had an altitude and azimuth refractor, by Dolland, of ten feet focal length, and five inch object glass. This at the time of its reception in 1830 was much the largest refractor in the country. Its mounting in the tower of one of the college edifices was by no means equal to its optical power. Williams College had a Herschelian reflector of ten feet focal length, equatorially mounted and a very good transit instrument. Hudson Observatory, in Ohio, had an equatorial refractor of four inches aperture. The fine telescope of the Philadelphia Observatory — an equatorial of eight feet focal length and six inch object glass, by Merz and Mahler, of Munich — was received in the autumn of 1840. Its erection marks an era in the history of our astronomical instruments. In optical power, and in the fineness of its graduations, it was quite superior to any thing which we then had. It was, moreover, the first of the fine German telescopes imported into the country. When I have mentioned these instruments I have given a summary of our means of astronomical observation at the close of 1840. We had many other telescopes, that were useful in observing eclipses, occultations, and transits, and which afforded interesting views of many of the heavenly bodies, but none with high optical powers and accurate graduations, for the exact determination of the elements of position.

What have we done in this department within the last twenty years? I will not occupy your time with a history of American observatories, which must be familiar to the most of you. Suffice it to say, that we have now several not unworthy of comparison with some of the old and world-renowned observatories of Europe. Astronomical instruments have two objects in view; the one is the increase of optical power, the other is the exact determination of position. The one class of instruments scrutinizes the celestial bodies, and shows us *what* they are, the other tells us *where* they are.

As it respects optical power, we have at least six telescopes in use, which are entitled to high rank. The first in size is the great Munich Refractor at Harvard University, of twenty-two and a half feet focal length and fifteen inch object glass, made after the same models as the great refractor of the Russian observatory at Pulkova. These two instruments stand alone. Among refractors they have no equals. The results given by the Cambridge instrument in researches upon the nebulæ, upon the ring of Saturn, in the discovery of the eighth satellite of that planet, and upon the nucleus of the brilliant comet of 1858, fully attest its power. The Munich telescope at Ann Arbor, has an object glass of thirteen inches aperture. The telescope at the observatory of Hamilton College, in the State of New York, made by Spencer, has an object glass of thirteen and a half inches, and one at the private observatory of Mr. Rutherford, in the city of New York, made by Fitz, has also an object glass of thirteen and a half inches. These three instruments, two of them of American workmanship, are reputed to be of excellent quality. They are all equatorially mounted. The refractor of the Cincinnati Observatory, with a twelve inch object glass, and that of Washington Observatory, with an object glass of nine and six tenths inches aperture, have both rendered good service to astronomy, and have acquired a high reputation for excellence. They are both from Munich. The instrument at Washington is of the same size as that which was used by Struvè for many years, with such signal success, at Dorpat.

For the coördinates of position, the right ascension and declination, we have mounted and in use, six meridian circles of the most refined construction, reading off to a single second of arc, and with the microscopes attached, descending practically to tenths of seconds, while the chronometers give the time to intervals much less than the tenths of seconds. These circles are at the observatories of Cambridge, Washington, Georgetown, West Point, Ann Arbor, and Albany. The last-named instrument, constructed under the special supervision of the late director of the Dudley Observatory, Dr. B. A. Gould, contains improvements which, it is believed, place it second to no other of the kind which has ever been made. It will remain a proud monument of the ability and practical skill of the director. It is understood that the other instrumental equipments of this observatory will be of the highest order. And it is hoped that an institution so nobly conceived and so auspiciously commenced, will, notwithstanding the adversity which has unhappily befallen it, and checked its progress, yet fulfil the distinguished promise of its inauguration.

In addition to these, there are in the possession of collegiate institutions and in the hands of amateur astronomers, I think, not less than twenty instruments, which are capable, under favorable circumstances, of showing the principal division in the ring of Saturn, and of opening double stars from 1″ to 2″ apart. Many of these are equatorially mounted, and from time to time will be able to furnish most valuable contributions to astronomy.

It will thus be seen, that with our public appreciation of science, and the princely munificence of private citizens, we have in the short space of twenty years furnished ourselves with the means of embarking with advantage upon the most refined and difficult branches of practical astronomy. In addition to this, we may say, without undue self-esteem, that we have many observers among us who have attained to great practical skill in observation; observers who have studied the theory of errors in construction and adjustment and observation, and have made themselves proficients in the best method of detecting and eliminating them. The published proceedings of this Association bear honorable testimony to the profound ability with which the most difficult problems of celestial mechanics have been discussed before us. The volumes of the Washington and the Cambridge Observatories speak for themselves. They show that American observatories may be relied upon for the most accurate determinations. The astronomical expedition to Chili, which had for its object a new and independent determination of the solar parallax, reflects great credit upon the scientific enterprise and ability of Lieut. Gillis. The elaborate computations made by Dr. Gould, give for that important element a result extremely near to the received value as based upon the transit of Venus in 1769. And, what is pertinent to our present purpose, the new determination (8″. 5) is based alone upon American observations.

It is understood that the observatories of Cincinnati and West Point contain records of high value which have not been made public. The astronomical and geodetic labors of the United States Coast Survey, including the method of longitudes by the electric telegraph, show consummate skill in the conduct of difficult researches, and reflect the highest honor upon the scientific character of our country. We think it may be said with safety, that none of the great geodetic enterprises of modern times, beginning with Picard's, has been more thoroughly accurate than our own. The application of the electric telegraph to transit observations, known among the astronomers of Europe as "The American Method"—because conceived and perfected by American genius—is undoubtedly one of the greatest improvements in practical astronomy known to the history of the science. It enables the observer to do in a given time quadruple the work which was possible without it, and with nearly quadruple accuracy. The American Nautical Almanac, representing numerically the latest and highest improvements in planetary astronomy, shows what progress we have made in the difficult art of computation. I call it a difficult art because in a work of five or six hundred pages of closely printed figures, the intention and the expectation is that it shall be free from errors. To achieve this is a work of the utmost difficulty. It is a record of the ability of American computers, and in some respects a record of American science. I may extend this last remark to the Astronomical Journal, edited by Dr. Gould, which, in the main, presents American astronomical science in its most refined, practical, and analytic forms.

In the manufacture of the requisite qualities of glass for telescopes and in the construction of instruments of the very largest descriptions, we have not yet attained to the excellence of our older neighbors on the other side of the Atlantic. But who shall say that even that is a forbidden field to the genius or American artisans? Who shall say that Guinard and Frauenhofer will not find their proud rivals in American workshops to share with them the glory of perfecting the noblest of instruments, the astronomical telescope? The eminent success of Spencer, Fitz, and Clark, in telescopes of large size, to say the least, gives us substantial ground to hope that with due encouragement on the part of the public, we shall erelong in this department be able to meet the severest demands of advancing science.

There is still another and higher department of astronomy which, in some directions, forms the solitary basis of further progress. Without it, facts may be observed and accumulated, but the remote and hidden laws of which they are consequences and ultimate exponents, can never be reached. I refer to that higher mathematical analysis which seizes the actual conditions of the system and demonstrates from them the laws which must exist, or which takes the laws as we now have them and demonstrates what must be the condition of the system at any time to come. A higher and nobler function of the human mind than this we can hardly conceive, below that of paying its devout homage to the Supreme Architect of these countless worlds above us. And here the name of Bowditch stands as a proud monument of American science. The profoundest problems of the Mecanique Celeste did not evade his searching scrutiny. But the name of Bowditch stands not alone. That of the lamented Walker is not unworthily associated with it. Other names there are, which delicacy forbids me to mention in this presence. Posterity, with even-handed justice and pious care, will assign them an honorable place among the great geometers of the age.

I have before alluded to the problem of the sun's distance. It is well known that the determination of that important element rests almost exclusively upon the observations on the transit of Venus in 1769. It is also known that the reliable observations on that occasion were fewer than could have been desired, and, strange to say, some which were given to the world were of doubtful authenticity. And we may add that however valuable the results obtained from observations on Mars in opposition, and Venus in conjunction, the next transit of Venus in 1874 is regarded as the surest means of verifying or correcting the previous determination of the sun's parallax. The astronomer royal of England has already called the attention of astronomers to this problem, and says that it is not too soon to begin the preparation for so important an event. He looks to the astronomers of the United States to take a leading part in the enterprise, and adverts to the peculiar advantage which will result from connecting together a series of stations covering a great extent of country by that wonderful auxiliary to science, the electric telegraph.

To all the foregoing, I may add that the application of the art of photography to astronomy, now regarded as another "wonderful auxiliary," is due to American genius. The late director of the Cambridge Observatory, the lamented Bond, was the first to make the application, and show its practical importance. That sun-painting art, which delineates so quickly and so truthfully the features of the "human face divine," delineates with equal facility and equal precision the features of the sun, moon, and planets. The relative position of double stars and groups of stars is given with an accuracy scarcely inferior to the best micrometric measurements.

From the facts which have now been stated, the question again recurs, *What may justly be expected from American astronomers in advancing their chosen science?* But the facts themselves emphatically answer the question; — and that answer is, *Much every way.*

The material equipments are at hand, or if not at hand, will soon be furnished when solicited by the spirit of devotion and earnest progress. True, we have not the princely and governmental patronage which gives vitality and strength to many scientific establishments of Europe. But we have what few other countries can boast to the same extent, an enlightened public sentiment, a just appreciation of the dignity and importance of science, penetrating every portion of our community, and commanding not only the wealth of merchant princes, but the wages of hardy industry, to achieve the object of scientific culture. With a dependence like this, which is elastic, buoyant, versatile, and which readily adapts itself to the exigencies of pressure and demand, we need not shrink from the noblest enterprise. We have skilful observers, we have ready and practised computers, we have mathematicians who will be responsible for any contributions which analysis can make to the common progress. Such, gentlemen, is our position; such is the responsibility which rests upon us.

It remains for us to aid and encourage each other and show what may be done by a generous devotion to a noble cause. It may be that some of us will have occasion to profit by the precept of the Hebrew preacher, "If the iron be blunt and he do not whet the edge, then must he put to more strength." With the strength of firm resolve and the whetted edge of diligence, we may enter with good hopes upon the task before us. Our united efforts will surely not be fruitless.

www.ingramcontent.com/pod-product-compliance
Lightning Source LLC
LaVergne TN
LVHW020741120826
845150LV00010B/2287